Center for Hydrogen Safety Conference 2019

Sacramento, California, USA
14 – 15 October 2019

ISBN: 978-1-7138-0575-5

Printed from e-media with permission by:

Curran Associates, Inc.
57 Morehouse Lane
Red Hook, NY 12571

Printed with permission by Curran Associates, Inc. (2020)

For permission requests, please contact AIChE
at the address below.

AIChE
120 Wall Street, FL 23
New York, NY 10005-4020

Phone: (800) 242-4363
Fax: (203) 775-5177

www.aiche.org

Additional copies of this publication are available from:

Curran Associates, Inc.
57 Morehouse Lane
Red Hook, NY 12571 USA
Phone: 845-758-0400
Fax: 845-758-2633
Email: curran@proceedings.com
Web: www.proceedings.com

TABLE OF CONTENTS

Hydrogen Storage for Refueling Stations

CENTER FOR Hydrogen SAFETY 2019 CONFERENCE

Authors:
Felbaum, J. W.

HYDROGEN STORAGE FOR REFUELING STATIONS

- HYDROGEN REFUELING DEMAND
- PRESSURE VESSEL TYPES & KEY ATTRIBUTES
- COMPARISON OF ASME SECTION VIII DIV 1, 2 & 3 VESSEL DESIGNS
- ASME DIV. 3 TYPE I 15,000 PSI MONOBLOC DESIGN FOR HYDROGEN STORAGE
- DESIGN TARGETS
- NOTCH SIZE IMPACT ON CYCLE LIFE
- SURFACE CONDITIONING
- HEAT TREATMENT OF MATERIAL
- SPECIAL POST HEAT TREATMENT HYDRAULIC TEST
- BASIS FOR K_{max} LIMIT
- CRACK GROWTH RATE IN HYDROGEN AND AIR
- 3-ZONE HYDROGEN CASCADE FUELING STATIONS ASME 1,034 BAR / 700 BAR / 500 BAR
- CYCLE COMPARISON WITH AND WITHOUT AN INTERMEDIATE 700 BAR ZONE

Overview and Recent Developments for the Hydrogen Risk Assessment Models (HyRAM) Toolkit

Authors:
Ehrhart, B.
Muna, A. B.
Sims, C.
Hecht, E. S., Sandia National Laboratories
Blaylock, M. L.
LaFleur, C. B., Sandia National Laboratories

Hydrogen Risk Assessment Models (HyRAM) is a software toolkit that provides a basis for quantitative risk assessment (QRA) and consequence modeling for hydrogen infrastructure and transportation systems. HyRAM integrates validated, numerical models of hydrogen behavior, statistics, and a standardized QRA approach to generate useful, repeatable data for the safety analysis of various hydrogen systems. This talk provides an overview of the models contained within HyRAM and how to use them to obtain relevant information regarding potential risk. Recent new features and changes to the software are also described, as well as future plans for new models and materials. HyRAM is a software developed by Sandia National Laboratories with support from the U.S. Department of Energy Fuel Cell Technologies Office. Sandia National Laboratories is a multimission laboratory managed and operated by National Technology & Engineering Solutions of Sandia, LLC, a wholly owned subsidiary of Honeywell International Inc., for the U.S. Department of Energy's National Nuclear Security Administration under contract DE-NA0003525.

Use of Active Monitoring for Hydrogen Facility Risk Reduction

Authors:
Buttner, W.
Wischmeyer, T.
Post, M.
Thorson, J.

Hydrogen sensors are a critical safety element that facilitate the safe implementation of the hydrogen infrastructure, and therefore are often required by code for many applications (e.g., repair garages, fueling stations). The NREL Sensor Laboratory was established by the DOE Fuel Cell Technology Office to promote the availability and use of sensors. The research activity of the Sensor Laboratory is evolving from its original mission of performance evaluation to support sensor developers and to verify metrological specifications for end-use applications. While sensor performance evaluations remain a core capability, the Sensor Laboratory is applying a more holistic approach on the role of sensors in promoting hydrogen safety through risk reduction. Towards this end, the Sensor Laboratory RD&D effort has expanded to include the development and deployment of a hydrogen wide area monitor (HyWAM) to elucidate the behavior of cold hydrogen plumes; the HyWAM has also been used to validate a hydrogen dispersion model of ambient temperature indoor hydrogen releases. The outcome of the hydrogen dispersion measurements is to support the validation of models that can be incorporated along with monitoring strategies into quantitative risk analysis of facility design and operations to lower risk. The goal is to make hydrogen monitoring an integral part of an integrated safety system approach to reducing risk at a hydrogen facility (such as a hydrogen fueling station with LH2 storage). In contrast to past presentations from the NREL sensor laboratory, in which the outcome of a specific research topic or topical study was presented, this talk will focus on the evolving research direction of the Sensor Laboratory of using sensors as an integral part of the safety system to actively contribute to the facility design for risk mitigation.

An Explosion in Ammonia Back End Process Caused by Missing Path of Maintenance Work

Authors:
Widodo, A. S.

The objective of this paper is to share the past undesirable experience in PT Pupuk Kalimantan Timur in order to be lesson learned on Hydrogen safety. In the beginning of 2018, in Ammonia plant Kaltim 2, there was an incident in back end process that lead to damage in plant assets. The initial problem was identified from leakage of SynGas, Hydrogen content is 74%, in one of heat exchanger in Back End Process. To minimize losing of SynGas to the environment and unsafe condition due to spreading of Hydrogen to the atmosphere, corrective action has been conducted immediately. The working preparation such as Job Safety Analysis (JSA), working procedure, contingency plant was done completely before the corrective work performed. Unfortunately, the explosion occurred while the working is in progress. Although there is no fatality, but Ammonia plant must be shut down and losing of some plant assets. The investigation has been conducted by independent team to find the main cause of the incident. The result of investigation report found that the key problem is a missing path of the previous maintenance work at Turn Around (TA) in 2017. Finally, based on this incident, the author concluded that all personnel in maintenance work should take into account every step of work appropriate with standard and procedure. Also, joint inspection among related parties can be conducted to ensure no missing step in maintenance program.

Keywords : Hydrogen, SynGas, Ammonia plant

Review: Hydrogen Tank Explosion in Gangneung, South Korea

Authors:
Kim, S. I.
Kim, Y.

A devastating hydrogen tank explosion occurred in Gangneung, South Korea on May of 2019. Two men died and several buildings including even for more than 100 meters away, have been seriously damaged. The cause of the explosion has been found that oxygen is permeated into the hydrogen storage tank. It was a P2G testing site and a hydrogen is produced by the water electrolysis where (3) hydrogen storage steel tanks of 40,000 liter (ID: 2,450 mm, H: 7,668 mm) with operating pressure of 1 MPa. The equivalent TNT is estimated to be of about 50 Kg based on the damage near by. There was no secondary fire and the damage is merely by the detonation pressure. Within 15 meter distance from the hydrogen tank, the structure beam of the building has been damaged and most of the windows of the 5 story building which is about 100 meter away and faced to the hydrogen tank, are all crushed. The origin of the ignition inside of the hydrogen tank has not been proved. The details of accident will be presented.

Optimal Safe Fueling of Hydrogen Powered Industrial Trucks

Authors:
Minas, C.
Minas, A.

Hydrogen fuel cell powered electric trucks are widely used in industry where more than 20000 systems are currently installed, achieving more than 100 million operating hours. The electric trucks are equipped with fuel cell power systems which incorporate a permanently mounted Hydrogen (H2) tank containing compressed hydrogen gas and enabling onboard fueling. The H2 tanks vary in capacity, ranging from 20 to 150 liters and are made of either steel (4130) or composite material (aluminum liner with carbon fiber overwrap). Fueling can be performed several times a day depending on the usage of the electric truck. It is critical to develop optimal fueling designs in order to minimize the fueling time and maximize operator productivity without exceeding the maximum allowable temperature of the H2 tank. A dynamic model was developed which included transient heat convection, conduction and fluid flow to predict the temperature change of the gas in the fueled tank using a finite difference approach. The input variables include ambient temperature, supply temperature, supply pressure, initial tank temperature, specific tank qualities (tank geometry and material properties), and Dispenser orifice size. The output variables of the code include the fueling time, fill ratio, the pressure and mass of the supply reservoir and fueled tank vs. time and the temperature of the H2 inside the fueled tank and the temperature distribution through the tank wall thickness vs. time during the fueling process. The fill ratio is defined as the final mass in the tank over the max allowed mass in the tank defined by the governing standard (SAE J2601-3). The model was validated using experimental data of various H2 tanks. The model was used to determine optimal orifice size for fleets of different H2 tank sizes and types. A multi-orifice design was also developed which optimized the fueling process.

Hydrogen Application in Maritime Industry: Safety Codes & Standards Requirements

Authors:
Pal, N.
Pratt, J.

Hydrogen application in maritime industry: Safety codes & standards requirements

Dr. Narendra Pal and Dr. Joseph Pratt

Golden Gate Zero Emission Marine, Alameda, CA

With present contribution of GHG emissions from international shipping, there is an urgent need of annual GHG reduction considering vessel population growth which largely off sets efficiency improvements. On a pathway of CO_2 emissions reduction consistent with the Paris Agreement temperature goals, IMO has set aim to reduce the total annual GHG emissions by at least 50% by 2050 compared to 2008.

To achieve IMO's goal of at least 50% annual GHG emissions reductions by 2050, we need a new transportation technology with emissions reductions approaching 80% or more, to make the emission reductions robust against growth in either population, or growth in the intensity with which technology uses energy.

Zero-GHG energy pathway is possible with hydrogen. High efficiency hydrogen energy conversion devices such as fuel cell that converts hydrogen into electrical power are powerful drivers for hydrogen technology. Today commercially-useful, zero emission vessel powered by hydrogen is possible to build.

The existing maritime regulations have been formulated to cover liquid natural gas (LNG) as a primary propulsion fuel. All the ships using gases or other low flash point fuels have to comply with the International Code of Safety for Ships using Gases or other Low-flashpoint Fuels (IGF Code). In order to ensure the safe and timely deployment of hydrogen technologies in maritime, there is a need for developing / adopting code for hydrogen application in maritime in consultation with other existing codes such as National Fire Protection Association Code on Hydrogen Technologies (NFPA-2) considering harsh environmental conditions and associated safety for maritime vs on-land use of hydrogen.

Abating Volatile Organic Compounds in Low Pressure (ATMOSPHERIC) Hydrogen Process Flows Using a Regenerative Thermal Oxidizer

Authors:
Addagarla, S.
Miller, P.

Thermal oxidizers are a well-known technology for the destruction of volatile organic compounds (VOC). Regenerative thermal oxidizers (RTO), an EPA referenced technology for VOC abatement, was found to be a very attractive option to abate VOC's in our laboratory environment.

Our laboratory conditions produced challenges in managing the operation of the RTO. The VOC's that needed to be abated were in a process flow that included up to 95% of hydrogen by volume. The laboratory test work, being variable in nature, resulted in process flows between 50 scfh and 1300 scfh, which the RTO needed to accommodate midstream during operation. These low process flows led to the design and building of a unique, small size, RTO to abate the VOC's.

The complexity of the process flow into the RTO was further compounded by the need to maintain vacuum conditions in the inlet line. Vacuum conditions increase the propensity for air leaking into the piping system, with the potential of air mixing with the hydrogen in the process line, creating combustion prone mixtures. Therefore, flame arrestors, monitoring equipment and alarms were put in place to ensure safety during operation. Vacuum conditions in the system impacts the fan/blower selection and the burner system for the RTO.

In addition to the operational conditions, we discuss the instruments and controls for the continuous monitoring of VOC's and Hydrogen. The availability of an instrument, to measure in real time, the LEL levels was not an off-the-shelf purchase and required much experimentation to identify a viable, long-term solution. To understand the stoichiometry of hydrogen-air in the region where the air is mixed with the process flow, we present the results of the computational modeling work that we did and the safety precautions we took with the piping in this mixing region.

Effective Application of QRA to Hydrogen Fueling Operations

Authors:
Vilas, K.
Shah, J., Baker Engineering and Risk Consultants, Inc.
Pierorazio, A.

This paper focuses on the challenges associated with assessing risk for hydrogen fueling operations. Currently, the commonly used method for assessing risks involves a PHA/LOPA approach to identify potential consequences and associated frequencies to determine scenario-based operational risks. In addition, the 'worst case consequence' event(s) is typically reviewed to determine the level of safeguards required for safe operation. However, this approach to quantifying hydrogen fueling operational risk fails to address the lower consequences, higher frequency events that may drive site risk profiles.

The hydrogen fueling industry has already experienced events, including a recent explosion in Norway that resulted from an improperly-installed plug on a high-pressure storage tank. While this incident resulted in limited damage, it called into question the safety of an up-and-coming technology. With a shift towards clean energy and the projection that hydrogen production may be cost competitive before 2030, many companies are investing in this technology as part of their overall corporate portfolios. It is essential to define the right approach to safety in the initial hydrogen network design and implementation in order to mitigate the potential for a negative perception towards the technology.

A properly conducted risk analysis for hydrogen fueling systems should incorporate the best practice work currently being done through the PHA/LOPA process with more quantitative approaches to better understand the risk profile. This paper addresses how a comprehensive scenario-based approach that considers existing process safeguards, as well as new research in the explosion and ignition characteristics of hydrogen, can be utilized to determine quantitative risk profiles. The quantitative analysis enables optimization of station locations, station layouts, and selection of safeguards to provide decision-makers with an objective framework to manage risk exposure to a tolerable level. Additional challenges to consider will also be addressed to promote discussion and improve future planning in the US.

Novel Risk-Based Design Approach for Hydrogen Refueling Stations

Authors:
Ade, N., Texas A&M University
Wilhite, B., Texas A&M University

In order to move towards further commercialization of the Hydrogen economy, it is essential to minimize the associated safety concerns and maximize the profitability of the involved systems such as that of Hydrogen refueling stations. Currently, Hydrogen refueling stations are primarily designed to satisfy required consumer demand while ensuring that Hydrogen is supplied at a profitable rate (13-16 $/kg). The developed design is then supported by a quantitative risk analysis (QRA) carried as per the NFPA-2 standard to validate that the associated risks (such as that of fire and explosion) are acceptable. However, the absence of QRA in the design stage itself leads to a possibility of the refueling stations being over-designed from the perspective of safety. As a part of this study, we have developed a novel risk-based design approach for Hydrogen refueling stations that allows the design of these facilities with optimized cost, quality, and risk. The Hydrogen refueling station is designed using state of the art process simulation platform, SimCentral that allows integration of quantitative risk analysis. The results obtained from this study provide deeper insights into the risk-benefit counterbalance associated with the design of Hydrogen refueling stations and provides design guidelines to facilitate further commercialization of the Hydrogen economy.

AUTHOR INDEX